The Study of Life and

Vladimir I. Vernadsky

The Study of Life and the New Physics
l'Etude de la vie et la nouvelle physique

translated by Meghan Rouillard

21st Century Science Associates • Washington, DC

Copyright © 2015 by 21st Century Science Associates

All rights reserved. No part of this publication may be reproduced, stored in a retrieval system, or transmitted, in any form or by any means, electronic, mechanical, photocopying, recording, or otherwise, without the prior written permission of 21st Century Science Associates, P.O. Box 16285, Washington, DC 20041.

A Note on the Text

Vernadsky originally published this essay in French as "L'Etude de la vie et la nouvelle physique" in the 31 December 1930 issue of the *General Review of Pure and Applied Sciences* (*Revue générale des sciences pures et appliquées*) in Paris. A Russian version was published in the *Bulletin of the Academy of Sciences of the USSR* (*Известия Академии наук СССР*) in 1931. This translation is based on Vernadsky's original French publication.

Footnotes are translator's, unless otherwise indicated. The relatively small number of footnotes from the 1930 French edition, presumably included by Vernadsky, are in brackets. Emphasis is original, unless otherwise indicated.

Translator's Introduction

In this revolutionary piece, published in 1930 in French in the *Revue générale des sciences pures et appliquées*, Vladimir Vernadsky makes a powerful and provocative argument for the need to develop what he calls "a new physics," something he felt was clearly necessitated by the implications of the groundbreaking work of Louis Pasteur among few others, but also something that was required to free science from the long-lasting effects of the work of Isaac Newton, most notably.

For hundreds of years, science had developed in a direction which became increasingly detached from the breakthroughs made in the study of life and the natural sciences, detached even from human life itself, and committed reductionists and small-minded scientists were resolved to the fact that ultimately all would be reduced to "the old physics." The scientific revolution of Einstein was a step in the right direction, but here Vernadsky insists that there is more progress to be made. He makes a bold call for a new physics, taking into account, and fundamentally based upon, the striking anomalies of life and human life.

Meghan Rouillard

Vladimir Ivanovich Vernadsky (1863–1945)

The Study of Life and the New Physics
Vladimir I. Vernadsky

I.

The revolution which in our 20th Century is taking place in physics, raises, for scientific thought, the necessity of a new revision of fundamental biological representations. It seems that for the first time it is becoming possible, in the cosmos constructed by science, to promote the phenomena of life to an important place. It is for the first time in the course of three centuries that the possibility of overcoming the profound contradictions created by the historic progress of thought is opening to us: contradictions between the scientifically constructed cosmos and the life of humanity; between the conception of the surrounding world connected with the conscience of man and its scientific expression. This contradiction has penetrated our intellectual life since the 16th Century; we feel it profoundly with each step. Its consequences are innumerable.

It is therefore important to follow attentively and to contemplate the development of the new physics, as the changes produced in our life due to the creation of the new scientific picture of the Cosmos—a consequence of the new physics—in which the contradiction with the human sentiment will not exist, these changes grow with the progress of physics.

This revolution must have no less of an impact upon the essential instrument of scientific thought—everyday scientific work, the psychology of researchers, because it has created, as we will see, a striking unconformity in the course of recent centuries between the scientific picture of the world and the scientific work upon which it is based.

Thus we are witnessing one of the greatest processes in the progress of scientific thought, and one of the age-old crises of human consciousness.

II.

Our scientific picture of the cosmos has its genesis at the time of the Renaissance. In the 16th Century, Giordano Bruno (1548–1600) clearly expressed the infinity of the universe and the small place occupied by our Sun, not to the mention the Earth.[1] Nicholas of Cusa (1401–1464) had understood this and expressed it one century before him.[2] Bruno said, with greater clarity than others, something which, in that time, was raised in all areas of human consciousness. In fact, the construction of Bruno was not a scientific acquisition, but he drew unprecedented philosophical conclusions from new scientific discoveries, conclusions which surpassed that which was scientifically known, and which were in agreement with the later development of scientific understanding. The entire scientific conception of the universe changed in a radical way. The tradition of thousands of years was shattered.

The philosophical constructions deduced from new facts and scientific empirical generalizations got ahead of the later acquisitions of precise scientific thought by several generations.

Based upon the telescope, a new conception, a new sci-

---

1. Bruno, tried for heresy in 1593 by the Roman Inquisition, had posited that stars were suns with their own exoplanets, and that they could potentially harbor life. Vernadsky's particular reference here is to his insistence that the universe was infinite, and had no body at its center. To quote Bruno: "The universe is then one, infinite, immobile.... It is not capable of comprehension and therefore is endless and limitless, and to that extent infinite and indeterminable, and consequently immobile."

2. It should be noted that this reference to Cusa does not appear in the Russian Academy of Sciences publication of this piece.

Nicholas of Cusa (1401–1464), founder of modern science and leading organizer of the Renaissance, whom Vernadsky describes elsewhere as "one of the most original and prodigious minds of his time."

entific sense of the Universe[3] developed in the course of a small number of generations: in the course of several decades, Copernicus, Kepler, Galileo, and Newton shattered the age-old relationship that had been formed between

---

3. Capitalization is consistent with the original. This may vary throughout in the case of Universe, Cosmos, etc.

man and the universe.

The scientific picture of the Universe, embraced by the laws of Newton, left no room for any manifestation of life, despite its appearing to have reached the limits of scientific perfection.

Not only man, not only life, but our entire planet was lost in the infinity of the Cosmos. Up to now, man and, through him, the phenomena of life, occupied a central place in the Cosmos, in the scientific, philosophical, religious and artistic constructions; at the end of the 17th Century all of these representations disappeared from the scientific concepts about the Universe.

Ascribing excessive dimensions to the world, the new scientific concept of the Universe seemed, at the same time, to belittle man with his interests and conquests, to belittle all the phenomena of life, to the point of being a species of insignificant detail in the Cosmos.

It seemed that the more human thought developed, the more such a scientifically constructed Cosmos, totally foreign and inconceivable to all that lived, to every human personality and to human life, emerged with more vigor and clarity.

After Newton, this picture of the Universe, devoid of life, penetrated by scientific thought, more quickly established itself outside of any philosophical or religious representations, due to the scientific observation of surrounding nature.

Its importance was especially developed during the periods of great success of stellar astronomy.

The first of these epochs came at the turn of the 19th Century, the time of William Herschel and his sister Caroline Herschel, who discovered a new world and demonstrated for the first time the regularity of its construction, in particular the existence of an infinite number of nebulae, of stellar cosmic systems.

We are now living through the second epoch, in the 20th Century. The new blossoming of stellar astronomy is in large part due, on the one hand to powerful new methods of observation developed with an unprecedented momentum by the American observatories, and on the other hand to the immediate adoption of scientific observations by physics. The new astrophysical discoveries penetrate the new physics and are more and more guided by its constructions.

This is where the radical distinction of the new progress of stellar astronomy from those of the earlier scientific generalizations lies: from that of Hipparchus, Ptolemy, Brahe, the Herschels, and the Struves.[4]

Within the scientific milieu and that of learned men, in the 18th and 20th Centuries, voices were immediately and unceasingly raised, which indicated with concern the futility of life, as well as all other great human desires, a futility which seemed to result from the grandiose picture of the Cosmos. These spiritual[5] and intellectual tendencies found their justification in the cosmogonies based upon these observations. The English astronomer M. Jeans pre-

---

4. The Struve family was a dynasty of five generations of astronomers from the 18th–20th Centuries in Germany and Russia, including most notably Friedrich Georg Wilhelm von Struve, who worked on problems such as double stars, Otto Wilhelm von Struve, who headed the Pulkovo Observatory, and his descendants, Ludwig Struve, Hermann Struve, Georg Otto Hermann Struve, and Otto Struve.

5. Vernadsky used the adjectives *"spirituel"* in the French and *"dukhovny"* in the Russian versions of this article. Their sense is often broader than most modern English usage of "spiritual," encompassing what English conveys by "mental," "intellectual," or "of the mind," as well. Henceforth this will be translated simply as "spiritual."

sented them again recently in his speeches which drew the attention of the entire world. The fragility and the insignificance of life, its accidental nature in the Cosmos, always seems to find new confirmations due to the progress of exact science.

But this new development of the scientific picture of the Universe, established in the framework of scientific thought, will today intersect for the first time another, more profound current of conception of the world, which changed the empirically obtained picture of the Cosmos in a radical way.

It is neither philosophical analysis, nor religious sentiment, but scientific thought which begins to introduce corrections, to shed light, in a new way, upon the scientific picture of the Cosmos, considered for a long time as a stranger to human life.

Based upon generalizations and astrophysical theories, this picture changes, unexpectedly for contemporaries, thanks to the influence of the profound revolution which the fundamental constructions of physics have undergone.

A new wave in the new scientific structure of the Universe is rising. It puts these burning contradictions which have existed for centuries into a new framework.

III.

Until now, man could only resolve the contradictions which existed between his own conception of the world and that of the scientific picture by addressing it through either philosophy or religion.

Over the course of many centuries, the scientist had not reconciled the fact that neither he, nor any living thing—consciousness, thought, intelligence—all there is which is higher for him, could in any way impact the scientific picture of the Cosmos, nor introduce corrections in the construction of the Cosmos,—created by science, except by borrowing from other domains of the spiritual life of humanity, those of philosophy, religion, and in part, art.

Remaining on the ground of the scientific conception, he had to accept the scientific picture of the Cosmos, foreign to life, and to treat as an error and an illusion the importance which he always gave in life to intelligence and consciousness, to all living things of which he himself was a part.

Faced with the impossibility of actually scientifically reducing phenomena of life to physico-chemical phenomena, taking as a basis the picture of the Cosmos of recent times, he brought about a great movement in the scientific environment and that of educated men, which proclaimed that sooner or later it would be done, without radically changing the founding principles which were considered as unshakeable.

It was estimated that intelligence, consciousness, the most elevated properties of life, should be able to be reduced, along with all the other physiological processes, to physico-chemical processes which are a part of the structure of the Cosmos. It was thought that all philosophical,

The Study of Life and the New Physics • 9

Sir Isaac Newton (1642–1726), who sought to reduce all phenomena in the universe to a set of physico-chemical processes, which, according to Vernadsky, "left no room for any manifestation of life" nor could they be successful "in scientifically explaining consciousness, intelligence, and logical thought." Vernadsky therefore demands that science be liberated from the reductionist beliefs of Isaac Newton and that a "new physics" and a "new scientific picture of the Cosmos" be established on the basis of the study of life and human life.

artistic and religious manifestations of human consciousness would be included in Newton's scientific framework of the Universe without exception.

Philosophical thought never reconciled itself with such a representation: the analysis of philosophers and a great number of scientists who had reflected on the founding principles of their knowledge had arrived at the conclusion that this representation did not flow from scientific knowledge, and that [belief in this reduction and inclusion][6] was essentially nothing but faith, which was based on philosophical and even metaphysical representations.

Philosophical admissions, foreign to the exact sciences, constitute the basis of another tentative scientific explanation, having as its objective to become the master of contradictions, the acceptance of forces or forms of energy or of entelechy specific to the phenomena of life, foreign to the inanimate world.

These vitalist representations were not able to enter into scientific thought in a lasting way, as their roots are not found in exact and empirical material of scientific generalizations and facts, but were introduced into science by constructions and foreign philosophical research.

Based only on the analysis of the fundamental content of science, scientific facts and the generalizations deduced from them, and relying only upon them, the scientist was forced to admit that there was not a real basis for the belief that the physico-chemical phenomena of Newton's picture of the Universe were sufficiently profound and vast to embrace all the phenomena of life, and that at the same time it was impossible to deduce from them, from their empirical material, vitalist representations which would have completed the picture of the Universe.

---

6. Added for sake of clarity.

Aside from the logical analysis of scientific knowledge and the scientifically constructed universe, it is the observation of the history of scientific knowledge of recent centuries which had to give him this conviction.

In reality, the explanation of life given by the models of the dominant conception of the scientific universe has not made progress in the course of recent centuries. The same abyss stands between living matter and non-living, abiotic matter, as during the time of Newton.

The models and the constructions of physico-chemical systems of the Cosmos of Newton have not, up to this point, succeeded in scientifically explaining consciousness, intelligence, and logical thought.

The scientist had to search for a way out of these contradictions, either in philosophical or religious thought, or in the reconstruction of the scientific Universe, in which the phenomena of life expressed in scientific facts and empirical generalizations had to be included, along with other manifestations of reality.

IV.

Despite the generally widespread conviction of the immutability of the modern scientific representation of the Universe, despite its very enhanced precision in the last century, this representation did not acquire, in its basic principles, either the sufficient resilience or authority such that the place which life found there could be considered as proven, and that the scientist, remaining only on the ground of scientific knowledge, had to swallow his pride, to submit to and recognize the futility and the insignificance of life in the Cosmos.

Religious and philosophical thought gave an entirely different place to life in the Universe. Philosophical research developed quickly in the course of three centuries (and what a development it was!) in the opposite direction to that of the scientific picture of the world, while the religious constructions incessantly changed the elements which collided with scientific thought.

The awareness of the phenomena of life and of their immense importance in the Cosmos was simultaneously deepened in philosophy, in religious creation, and in the life of humanity.

The evolution of scientific thought, in this spiritual environment, little by little, and imperceptibly for contemporaries, ate away at the belief in the possibility of including the phenomena of life in the scientific picture of the universe without radically changing it.

But there is more. The change in this direction was inevitably prepared for a new phenomenon—the development and the structure of the scientific organization of humanity.

This is a matter of what follows.

With the progress of scientific work, after the brilliant success brought to the 18th and 19th Centuries by the descriptive natural sciences, and the penetration of precise scientific methods in the domain of the humanitarian sciences in the same centuries, the place occupied by the scientific picture of the Cosmos in scientific knowledge continuously decreased. Indeed, the scientific picture of the Cosmos was only completed by an ever-shrinking number of scientific researchers. An ever-growing part of the persistent work of humanity lost its connection with the scientifically created picture of the Universe.

The face of science has been completely transformed in the course of these two and a half centuries which followed the Principles of Natural Philosophy of Newton; entire sciences were created which had not existed during that time, and the overwhelming mass of these new sciences is related to the study of life and of humanity in particular.

It is not to be doubted that well over nine out of ten of all scientists work in domains of science which have no connection with the picture of the Cosmos, falsely considered to be a result of scientific work as a whole.

They are not at all interested in this picture and do not encounter it in the course of any of their scientific activity. Its change does not arise in the domain of their knowledge. They completely pass over it.[7]

---

7. The problem of over-specialization in science was something Vernadsky struggled with in his youth. In 1896, he wrote: "I feel that I am becoming a specialist, part of my interests are receding and although along with this the intensity of the work in a specialized area is strengthened[,] this is connected with a well known narrowing of the mind. In this respect my expedition to the Urals [which he was then visiting with his students] has done much for me and two roads have clearly opened before me—one, although little productive and partly dilettantism, at the same time forces the mind to work more intensely

This is demonstrated in a striking way in the history of the biological sciences of the 19th Century for example. The theory of the evolution of species which still plays such an important role in the conceptions of the last 70 years, and in the entire life of humanity, doesn't enter into the scientific picture of the Cosmos, since life is not represented there.

The history of the theory of evolution has not yet been written from this point of view but it is very curious and produces a completely different effect upon us today which it did not have at the time upon the people who had been a part of its creation. It animated the cosmogonic evolutionary representations, but finds itself in strong opposition to the physico-chemical researches of biology. Its agreement with the Cosmos of Newton, that is, the possibility of completely reducing it to the physico-chemical principles forming the basis of the Cosmos, always seemed doubtful: perhaps more doubtful at the time of C. Darwin than in the period that followed. In all cases it exerted a large influence upon scientific thought and did not figure into the scientific picture of the Universe.

We are now at a turning point. It is possible that the unconscious progress of scientific thought of the last decades moved in a direction which has destroyed the belief in the possibility of reducing the phenomena of life to the parameters of the Cosmos of Newton.

---

and more broadly, the other is more productive, more defined—but at the same time confines the mind within the specific parameters and inevitably shrinks the horizon, placing a person within the ranks of scientific workers but not among the creators of the unfolding process." (Bailes, Kendall, *Science and Russian Culture in An Age of Revolutions: V.I. Vernadsky and His Scientific School, 1863–1945*, Indiana University Press, 1990, p. 68.)

V.

The ground was therefore prepared unconsciously in the psychology of scientific workers, in part by following the progress of the theory of evolution, as we will now see.

Science is not an abstract entity, self-sufficient, with an independent existence. It is a creation of human life and exists only within this life. Its content is not limited by scientific theories, by hypotheses, or by models of the picture of the universe created by them. This content is principally made up of scientific facts and their empirical generalizations. The real content of science is the scientific work of living individuals.

These living individuals, scientific workers, constitute science as a social phenomenon: their spiritual disposition, their mastery, the level of their understanding and their satisfaction with the work they have accomplished, their will—this global scientific viewpoint—are essential factors in the historic progress of scientific knowledge.

Science is a complex social creation of humanity, unique and incomparable to anything else. It has a much more universal character than literature or art, and has little relationship to the forms of life of the state and society. It is a global social formation, as the forces of facts and generalizations, equally obligatory for the entire world, form its base.

There exists nothing comparable in any other spiritual domain of human life.

Science is made up of living personalities, bound by this universal obligation. This is why it is by no means a matter of indifference if the fundamental theoretical *results* of their work are foreign to and have no relation with the scientific *work* of the overwhelming number of living personalities and thinkers who represent science.[8]

We are seeing this in the current period. The content of scientific work is, for the most part, not even reflected in the scientific picture of nature.

This can only continue because of a belief that the scientific work of scientists will end up being bound by the current scientific picture of the universe and will not contradict it, this faith still persists. Many people are expecting it, occupying themselves with their special work, and are not concerned about the future.

If the faith disappears, the contradiction between the content of science and the result of its work will arise for the investigators and will require resolution.

Collectively, scientists cannot reconcile themselves with the religious or philosophical solution to the contradiction. They will seek a scientific solution.

---

8. Emphasis added to clarify meaning.

VI.

Science is a singular unity and all domains of its expertise are, without exception, tightly connected. This empirical generalization is so rigorous that it cannot be changed by individual will.

There is more. We can say, in borrowing the comparison from another domain of human life, that science is profoundly democratic. All the work performed in the realm of science is fundamentally equivalent, for *sub specie aternitatis* [*from the perspective of the eternal*] science contains nothing of importance, nor of unimportance: its efforts all lead to the same, unique scientific character, to the unique—and obligatory for all without exception—scientific comprehension of the surrounding environment.

This conviction guides, in the most profound and inevitable fashion, all scientific workers.

But the belief in what scientific work is produced by the majority of scientific investigators, that the phenomena having to do with the study of life will finally arrive at penetrating into the scientific picture of the universe, without producing fundamental changes in it—this faith irrevocably evokes, in the opinions of the scientist, a value quite different in different domains of scientific knowledge.

This results in an acute instability in the scientific organization of humanity.

The primary admission that, by their essence, the mathematical, astronomical and physical-chemical sciences alone exercise an action on the comprehension of the fundamental bases of the present scientific picture of the universe—space, time, matter, energy—this admission which has often been expressed, but which has never really penetrated into the scientific environment, cannot be durable.

It cannot be, as a result of the ever-growing number of workers occupied by the study of living phenomena, owing to the results of their scientific work acquiring an increasingly strong influence on scientific thought, and their work exceeding the value, for scientific thought, of constructions of the scientific picture of the Cosmos. The history of the evolutionary ideas of the preceding century, which I have already pointed out, is instructive from this point of view.

Doubts are raised among naturalists, preventing them from admitting the primacy of the mathematical, astronomical, and physical-chemical sciences, a primacy inspired by the modern edifice of the scientific universe.

Two conclusions must inevitably give rise to the doubts of the empirical naturalist:

Cannot the life sciences effectively change the fundamental representations of the scientific universe—the representations of space, time, energy, matter—in a radical way? And is this list of fundamental elements of our scientific thought complete?

Can the naturalist seriously admit that the intelligence of *Homo sapiens faber* is the final manifestation of the evolution of species, the maximum of spiritual acquisition of organized beings? Or indeed, must it be believed that only the transitory spiritual possibilities of life are manifesting themselves before us on the Earth in the present geological epoch, and that there exist higher manifestations in this domain in some point of the Cosmos?

Without a negative response from science to these questions which inevitably arise, faith in the reality of the contemporary picture of the universe can include only a

The Study of Life and the New Physics • 19

relatively limited number of scientific workers.

What's more, scientists do not inhabit an isolated island. Great creative work of humanity takes place all

photo credit: Esther M. Zimmer Lederberg Memorial Website

Alexander I. Oparin (1894–1980), a contemporary of Vernadsky, was more of a political tool and opportunist than a serious scientist. Oparin, pictured here on the right during a visit in 1969 to NASA Ames Research Center, made it his life's work to argue the opposite point as that which Vernadsky sets forth in this paper, insisting that the origin of life, and even that of human cognition, could ultimately be explained by purely random reactions among non-living chemical building blocks. Oparin's fallacies continue to corrupt science today (See article: Rouillard, Meghan, "A.I. Oparin: Fraud, Fallacy, or Both," 21st Century Science & Technology, *Spring 2013, p. 42)* around them—fertile in many respects—in other spiritual domains, in religion and above all in philosophy, work which is absolutely contrary to the scientific conception created in the last centuries.

All this widens the contradiction which exists between scientific work and its fundamental, official result.

At present, the scientific organization of humanity lacks necessary stability and the result of scientific work is increasingly dissociated from its *content*[9] in the consciousness of scientists, whose numbers always increase.

---

9. Emphasis added for clarity.

VII.

Once such an instability in the essential instrument of scientific knowledge is recognized, this cannot continue.

This state of affairs has begun to change suddenly in the past decade following a new first-order event—the radical change of the physical sciences, in part, astronomical.

Space, time, matter and energy are clearly distinguished for the naturalist of the year 1929, from the space, time, matter and energy of the naturalist of 1900.

They are not only different; it is obvious that they cannot serve the scientific construction of the Cosmos, even under the clearly changed form under which they are currently manifest. New ideas are penetrating physics which draw the required attention of physicists to the phenomena of life. As it happens, these new ideas are expressed with greater clarity in the phenomena of life than in the ordinary objects of physical investigation. These traits, these elements of the construction neglected in the scientific picture of the universe, which change its Newtonian form, clearly cannot be grasped or studied until we introduce in some form the sciences of life into the picture of the Universe.

It is at the same time curious that the traits of life which had drawn little attention from biologists, today serve as the first-order phenomena of life.

It seems to me that the profound and growing change which is occurring in the sciences of life under the influence of the crisis in physics is becoming clear in that way.

Before moving on to the problem of fundamental conceptions of life, now demanding attention and precision in connection with this crisis taking place in the historic progress of the physical sciences, I will say a few words about the characteristic traits of this crisis.

VIII.

As I clearly cannot dwell upon the changes taking place before our eyes in the fundamental notions of physics in any detail, I will only concern myself with some problems in the historical process which is unfolding, problems which will be necessary for me in the later report.

What is essential is the complete change in our notions of space, time, energy, gravitation, and matter. The force of universal gravitation, acting instantaneously upon every considerable distance, has disappeared without a trace from our thoughts. Space and time are inseparable, and to understand physical phenomena we are forced to geometrically employ space of not three, but of four dimensions. The boundary separating energy from matter fades. Energy is propagated in strictly determined jumps—quanta.

The reversal of opinions and representations was produced with great celerity, and was quite unstable. The physicists still thought, at the beginning of this century, much differently than we do today. I remember a conversation I had more than 20 years ago with P. N. Lebedev, the eminent Russian physicist, who told me he was only secure speaking about the ether. This was at the time when the notion of the electron began to enter into physics. Currently, physicists try not to speak of the ether and there are some who doubt its very existence.

At this time, at the beginning of the century, the dawn of dynamic representations of matter and of energy seemed to blossom along with the ether. Certain scientists of great scholarship, possessing a philosophical erudition, such as, for example, W. Ostwald senior, considered the atomistic representation of matter to have been definitively buried.

There was an attempt to rid chemistry of it (Wald).[10] It happens that the contemporaries had not understood the process of scientific thought which was developed with their participation.

In two or three years, the atomic representation achieved an unprecedented success, and became dominant.

From then, it was only one or two years before we often heard affirmation that as of now the existence of the atom had really been proven and that the atomic theory of matter was no longer a theory, but a natural phenomenon which could be detected. The atomic theory of Bohr and Rutherford seemed to reign definitively. However, this reign came to an end. Today, the atom begins to fade from our minds: we speak of the wave theory of matter, on the one hand, and on the other, the impossibility of reducing phenomena to the motion of points in the areas of physics which treat the physics of the atom, of even smaller particles. The more exact the determination of speed of motion of particle becomes, the less exact will be the determination of its geometric position. Mechanical laws of the motion of points cannot be applied to these phenomena with sufficient precision.

The old dynamic representations are reborn under a new form, as foreign to the old as atomic physics of the 20th Century is to that of Gassendi.[11]

The change which opinions have undergone is very abrupt: there is no longer any established stability there; we will probably live for a long time in the fermentation of ideas which characterize the current state of physics. It

---

10. Vernadsky is likely referring to chemist František Wald (1861–1930): http://www.hyle.org/journal/issues/13-1/bio_ruthenberg.pdf

11. A reference to Pierre Gassendi (1592–1655), a French scientist and astronomer known as an early proponent of the atomic theory of matter.

The revolutionary scientific discoveries of Albert Einstein (1879–1955) and Max Planck (1858–1947) represented the first major steps in the 20th century to overthrowing the old physics, and emphatically the ideas of Newton in the case of Einstein. Here, Planck and Einstein are pictured in Berlin with the German physicist Walther Nernst in 1931.

is precisely this fermentation which will have influence upon the neighboring sciences.

There had been no room for irreversible processes in the physico-chemical phenomena encompassed by scientific theory in the picture of the Newtonian universe, which had reigned, at the beginning of this century. All natural processes in that context were always considered de facto as reversible. This principle constitutes the basis of the scientific representation of a Cosmos of the 19th Century. In the case where they seemed to be irreversible, only an apparent irreversibility was assumed and the idea of a very slow development —to the point of being absurd—of a reversible process was accepted, which usually allowed for the management of any difficulties created by experiment and observation relatively well. Today, the irreversible process plays another role in physics—probably a very important role. This admission is of great significance for the problems which occupy us. All the conclusions have not yet been drawn. It is possible that the irreversible processes are dominant in the Universe, as they seem to constitute the essence of phenomena in molecular physics, in the physics of microscopic phenomena, in the phenomena of heat and radiant energy, and of light.

No less important is the distinction between statistical laws and laws treating the elements of physical processes themselves. I already mentioned the atoms which correspond to them and the characteristics of the application of the laws of motion of points to these.

It is a phenomenon common to all processes located in the internal structure of the Universe—molecular or microscopic according to the modern expression—for regions where the hypothetical universal gravitation has never been able to penetrate.

Here is a case where the law of causality, in the regular sense, seems to cease to be applicable or is no longer present. This law of causality is the alpha and omega of the picture of the Newtonian universe. The idea on which it is based is clearly expressed by Laplace in his acceptance of the possibility of encompassing the Universe in a unique formula whose solution allows for the calculation of the motions of the planets, the development of thought, the motion of reeds, and the change of state of spiral nebulae.[12] Such determinism disappeared for a determined category of physical phenomena in modern physics. It is not insignificant that for these cases some physicists saw that there was not only an analogy with the biological individual, but a phenomenon of the same logical category. In the best case the unpredictable coefficients, from a quantitative point of view, will become part of the classical formula of Laplace.

There is nothing great or small in nature. If we admit such a disparity in the action of causality—for example of the impossibility of expressing every-

Pierre-Simon Laplace (1749–1827), whose theory of deterministic causality Vernadsky ridicules as accepting "the possibility of encompassing the Universe in a single formula whose solution allows for the calculation of the motions of the planets, the development of thought the motion of reeds, and the change of state of spiral nebulae."

---

12. Known as "LaPlace's Demon." From the horse's mouth: "An intellect which at a certain moment would know all forces that set nature in motion, and all positions of all items of which nature is composed, if this intellect were also vast enough to submit these data to analysis, it would embrace in a single formula the movements of the greatest bodies of the universe and those of the tiniest atom; for such an intellect nothing would be uncertain and the future just like the past would be present before its eyes." Pierre-Simon Laplace, *Essai philosophique sur les probabilités* (1814).

thing by the laws of motion—we will inevitably be forced to come to the same admission in other cases.

The analogies between the infinitely small of the molecular world and the grandiose bodies and spaces of the stellar world are numerous and real. It is always necessary to have this correction in view. The new physics today begins to accept, by the intermediary of its numerous representatives, the principle destroying at its very root the representation of the infinity of the Cosmos, which Bruno had caused to penetrate into the understanding of the Universe of modern times. The idea of the possible boundary of the Cosmos, the finiteness of its space, is beginning to enter into scientific representations under a new guise. Certainly the dimensions of this Cosmos are very vast. Its volume is no less than a radius equivalent to 10^{17}–10^{18} km, that is, say quintillions of kilometers, and the importance does not lie in the dimensions, but rather in the fact that the volume of the world has limits, that it is bounded. That is where its immense importance lies. In this way, we are actually approaching the Middle Ages of Dante with his bounded universe, more than the infinite space of the scientists of the 16th–19th Centuries.[13]

The change goes even further. We are clearly approaching the distinction between the space of physics and the space of geometry. The principle of symmetry is beginning to penetrate physics. We cannot, for example, understand in any other way the problem recently posed for purposes of experimental investigation regarding the speed of the propagation of light: is it identical in both

---

13. This concept seems to be most clearly represented in Dante Alighieri's *Divina Commedia*, which poetically describes nine celestial spheres of heaven, as seen in Michelino's fresco.

A fresco by Domenico di Michelino showing Dante Alighieri (1265–1321) holding a copy of his Divine Comedy, with its seven terraces of purgatory and nine spheres of heaven depicted in the background. Vernadsky refers to Dante's work as an early example of an idea of a bounded universe.

directions along the same line?[14]

Certainly all the new acquisitions and this boldness will not remain stable in science; they imply that the old Newtonian representation of the Universe had created a crack, its scientific certitude shaken and an infinite and ever-growing throng of new representations made due to this crack have created an opening allowing for a sudden burst all the more rapidly.

The scientific representation of the Universe based on universal gravitation and on the physico-chemical phe-

---

14. This could be a reference to Michelson-Morley type experiments, which continued into the 1920s and even later, designed to test for this kind of phenomenon.

nomena of which we have spoken and which people have thought about for three centuries—must break down.

The scientific picture of the Universe based on universal gravitation and on the possibility of scientifically expressing all ambient motion of particles by reversible processes, by a rigorous determinism calculated in advance, this picture changes and does not correspond to facts. The individual begins to penetrate into the world of physical phenomena.

The elements of the Cosmos which constitute its existence, considered at the level of a microscopic cross-section, have, it is possible, profound analogies with living individuals and organisms.

The order of Nature is other than had been believed. To reduce the entire environment to what had been conceived, was found, in the final analysis, to be too simplified and approximate.

IX.

This radical change of fundamental physical representations must inevitably have a clear impact on the position of living phenomena in the edifice of the scientific Universe, as a great number of admissions of the new physics are in no way expressed with as much clarity as in life phenomena: such as, for example, the irreversible character, in time, of physico-chemical processes observed in living organisms. The irreversible cycle in the time of the phenomena characterize life to a degree unknown in the crude nature which surrounds us. The irreversibility characterizes the life of the individual, and is clearly expressed for us in its death. The irreversibility is no less clearly expressed in the process of the evolution of species in the course of geological time.

The irreversible process of evolution, its direction determined in the same unique sense, can be followed from the Algonquin Era[15] until today.

It was certainly known for a long time, but not much importance was attributed to it, although its contradiction with the assertion of the possibility of reducing the phenomena of life to physico-chemical processes accepted in the Newtonian universe was understood. It is a very common manifestation of a lack of breadth of our logical analysis in the domain of scientific thought: perhaps inevitable given the complexity of the Cosmos and the weakness of the scientific instrument which serves us in penetrating into the unknown.

The phenomena of life, of radioactivity, of the interior of stars, are probably the most clear manifestations of ir-

---

15. More commonly known today as the Pre-Cambrian Era.

reversible processes in surrounding nature. This type of process finds its most distinct expression in the phenomena of life.

But this expression in living processes, which is so clear, of a physical phenomenon of nature which is absolutely cosmic, is not accidental or unique. The same observed fact is seen in the properties of space and it can also be indicated in energetic processes, in the properties of matter which construct living matter.

These consequences of life for the fundamental notions of the Order of the Universe compel us to introduce the phenomena of life into the universe of the new physics.

In the presence of the unity of all that lives, of life, we cannot know where the penetration of the scientifically constructed Cosmos by the phenomena of life will stop. In this respect, the future is probably full of big surprises.

It is necessary to approach this process, whose progress seems inevitable to me, in another way, in relying upon the scientific conceptions of life.

It is important to pay attention to the phenomena of life whose introduction in the domain of the scientific construction of the Universe is already beginning to become probable.

We are approaching a very rational epoch—and that of a radical change in our conception of the scientific Universe.

This change will not be, in its consequences, any less important than it was at the time of the creation of the Cosmos, based upon universal gravitation, and infinite time and space, the Cosmos penetrated by matter and energy.

This change will allow us to overcome the contradiction which exists between life and scientific creation on the one hand, and the scientifically constructed Cosmos on the other, a contradiction which is clearly apparent during the 16th–19th Centuries, the time of the creation and the development of the Newtonian Universe. This was, incidentally, the conception of the Universe of Newton without Newton, who had introduced to it the corrections of a believing Christian.[16]

The possibility of overcoming the contradiction while dwelling only within the boundaries of science, appears to be opening up to us today.

---

16. While it is difficult to definitively interpret this sentence, one can rightfully detect some sarcasm in Vernadsky's reference to Newton as a believing Christian. It should be noted that in the person of Newton, we find a mixture of both dead reductionism and wild "religious" speculation: Newton was known to have posited the idea that God should be able to intervene and wind up the Universe at his will to prevent it from running down, a point he debated through his intermediary Samuel Clarke with Wilhelm Gottfried Leibniz in the Leibniz-Clarke correspondence: "God Almighty wants to wind up his watch from time to time: otherwise it would cease to move. He had not, it seems, sufficient foresight to make it a perpetual motion."

X.

There is no doubt that life in the scientific picture of the Universe will appear to us in an unexpected form. All phenomena studied in physics and chemistry manifest themselves there in another form than that which they present to our sense organs.

Let us dwell on several phenomena of life, which at this moment require attention, due to changes taking place in physics.

I am not a biologist and I consider phenomena of life from another point of view than that which is customary in biology—their action on the cosmic environment of their life. Claude Bernard, one of the greatest biologists of the past century, always employed this expression—cosmic environment—when speaking of life. He clearly understood that life is not an insignificant terrestrial phenomenon, but a cosmic manifestation.

Many manifestations of life in this domain, quite worthy of attention, can be noted, some of which take on a planetary character, connected with the Earth, whereas others clearly exceed the limits of planetary existence, indicating the more general situation of life in the Cosmos.

Among the notable planetary properties of of life are:

1) Living matter is created and maintained on our planet by the cosmic energy of the Sun. There, it forms an integral part of the upper geosphere, the biosphere, an indissoluble part of its mechanism.

2) Solar energy is gradually transported by the intermediary of living matter into the deepest parts of the planet, its crust.

3) The quantity of matter in the biosphere penetrated by life is a constant value or almost permanent across geological time.[17]

4) Living matter enters, in the course of all geological time and in a uniform way, into the geochemical cycles of the chemical elements in the Earth's crust, playing a very important role there. In this way, living matter provides a determined geochemical energy in the migration of terrestrial chemical elements, energy whose primary source is the Sun.

5) Living matter is in a continual chemical exchange with the cosmic environment that surrounds it, but is never spontaneously generated there. In the course of all geological time, this living matter represents a unique unity, genetically linked, and clearly separated from this cosmic environment.

6) Biogenic geochemical energy tends toward its maximum manifestation in the biosphere (first biogeochemical principle).

7) In the course of the evolution of species, it is the organisms which augment, through their life, biogenic geochemical energy, which survive (the second biogeochemical principle).

---

17. Vernadsky seems to change his view later, in 1938. In "Problems of Biogeochemistry II" he states: "The mass of living matter of the biosphere is close to the limit and, evidently, remains a relatively constant value on the scale of historical time. It is determined, above all, by the radiant energy of the Sun, falling on the biosphere, and by the biogeochemical energy of the process of colonization of the planet. Evidently, the mass of living matter increases in the course of geological time, and the process of the occupation of the Earth's crust by living matter has not yet been completed."

8) In the course of the evolution of species, the chemical composition of living matter remains constant, but the biogenic geochemical energy provided by living matter in the cosmic environment increases.

9) With the appearance of man in the biosphere, conforming to the second biogeochemical principle, the action of life on our planet develops and changes by the effect of his intelligence to such an extent, that it becomes possible to speak of a special psychozoic epoch in the history of our planet, analogous to other geological epochs in the change effected in living nature on Earth, as during the Cambrian or Oligocene, for example. With the appearance of a living being on our planet gifted with intelligence, we pass into another stage of its history.

What is even more, here we clearly go beyond the limits of the planet, as everything indicates that the progress of the geochemical action of intelligence, of the life of civilized humanity, goes beyond the limits of the planet.

We see here a manifestation of life which, although being located on our planet, indicates properties of living things seemingly not bound by it. Let us note several of the most profound manifestations of life:

1) Human intelligence and the activity of life, organized by this intelligence, change the progress of natural processes and similarly they change the other manifestations of energy known to us, but in a new way.

2) This activity is regulated by the second biogeochemical principle, that is, that it tends towards the maximum manifestation [of biogeochemical energy].

3) We have never observed on Earth the formation of a living organism from abiotic matter without the partici-

pation of another organism (Redi's principle—irreversible process).

4) Organisms constitute autonomous systems which, in the cosmic environment, create volumes (thermodynamic fields) whose temperature and pressure are particular to them, distinguishing them from their environment.

5) Organisms can live in the environment of molecular forces, foreign to the laws of gravitation, as well as in the environment which these laws characterize. Their minimum dimensions reach 10^{-6} cm; they penetrate into the domain of molecules.

6) The smaller the organism, the more intense its geochemical energy, the more quickly it creates new organisms. The maximum speed of this creation (division) has determined limits. I will call it the *element of biological time*. Today, I will again return to this phenomenon.

7) The life of the organism is an irreversible process which ends sooner or later with death. All living matter which penetrates the biosphere is overall an irreversible process in geological time; in the succession of generations we observe neither the beginning nor the end of this process and it could be that they do not exist.

8) It is not a diminution of free energy, but a growth which is effected in the cosmic environment resulting from life. In this case, life proceeds in a manner contrary to the law of entropy. Very few other physical phenomena in the Cosmos are on par with life according to this point of view: as, for example, radioactive bodies. But the cause of this phenomenon in living matter is clearly different.

9) The thermodynamic field of the living organism possesses, contrary to the properties of the cosmic envi-

Francesco Redi (1626–1697), an Italian poet, naturalist, and physician who was first to enunciate the principle of "omne vivum viv," or "life comes from life," thus challenging the belief in spontaneous generation. This concept is referred to by Vernadsky as Redi's Principle.

ronment, a clearly expressed dissymmetry. We know of nothing similar for other natural bodies on Earth. The dissymmetry is expressed there as with the particular character of the symmetry of space, occupied by living matter, the existence of very clearly expressed polar, enantiomorphic vectors, but above all, by the pronounced lack of conformity which distinguishes the right-handed from that of the left-handed character of phenomena (Generalization of Pasteur).

10) The activity of organisms, at least that of its most developed forms, is not a purely mechanical process which could be calculated. This activity is individual and diverse for different individuals. The degree of its freedom of action is not clear, but it is different in each case and can always be established.

XI.

This list is not complete, but it indicates, with evidence, that life manifests itself in the Cosmos in other forms than those which biology normally displays.

What is important, from the point of view of the scientific picture of the Universe, is that the investigation of life indicates such traits of the structure of the Cosmos, which in other phenomena studied by science are completely lacking or are very weakly or indistinctly expressed. In that way only the study of life changes the scientific picture of the Cosmos, formed without its contribution, and reveals new traits about it. It changes, essentially, the representation of space, time, energy, and other fundamental elements of the structure of the world.

Here I will dwell upon two phenomena which will allow for the clarification of the important role which the investigation of life plays in the scientific picture of the Universe, created by the new physics: in particular the phenomenon of the dissymmetry of the space of living organisms and the phenomenon of biological time.

In the first case, this is a matter of new properties (a particular state of physical space), observed in living organisms, and in the second, new properties of physical time.

The dissymmetry of living matter was discovered 80 years ago—in 1848—by one of the greatest scientists of the past century, Louis Pasteur, who clarified its importance for the structure of the scientific Universe. Pasteur conceived of dissymmetry as a cosmic phenomenon and drew from it very important conclusions for the knowledge of life. His works must today draw the most diligent attention of the new physics. He returned several times to these ideas, always going deeper. He returned to them for the last time in a more developed form, in 1883, 46 years ago, and regretted not being able to treat them in greater depth experimentally; he considered this discovery as the most important work of his entire life, as the most profound penetration of his genius into the problems of science.

The fate of his ideas was peculiar: the main idea which Pasteur emphasized has not, even today, penetrated scientific thought. The public opinion of chemists considered its basis as doubtful.

It seems to me that this depends upon the fact that chemists never took into account, in all its breadth, the notion of dissymmetry, on which Pasteur relied, and that this notion had not been understood by his contemporaries.

It was submitted to a serious analysis by another brilliant Frenchman, Pierre Curie, in 1894. His formulation of ideas is exceptionally concise, which could make them appear abstract, but his main theorem—on dissymmetry—allows for no doubt and is clear in its concrete importance for the naturalist. It states:

> The elements of symmetry of causes must be found in the effects, and the elements of dissymmetry of effects must be found in the causes.

The principle of Curie irrevocably resolves the dispute in favor of Pasteur, regarding his statements which call for research into the cause of the dissymmetry of natural bodies in life phenomena.

The fate of the works of Curie, were, in this area, analogous to that of Pasteur. Prevented from continuing this work due to the discovery of radioactivity, he returned,

The Study of Life and the New Physics • 41

Louis Pasteur (1822–1895), whose unique discoveries concerning the characteristic dissymmetries of living matter informed much of Vernadsky's work, along with the discoveries of Pierre Curie. Vernadsky states: "The weeds of oblivion have covered the path tread by Pasteur and Curie. It seems to me that it is precisely by that path that the current wave of scientific work must now continue forward."

before his death in 1906, 23 years ago, to works on symmetry; judging from his journal notes, he had arrived at great generalizations in this domain. After his death—he was crushed by a cart in the streets of Paris—nobody picked up the thread he left, which slipped away from later physical analysis: the principle of symmetry, an analysis which concerns us in particular today.

The weeds of oblivion have covered the path tread by

Pasteur and Curie. It seems to me that it is precisely by that path that the current wave of scientific work must now continue forward.

It has been six years since the eminent Dutch chemist F. Jaeger, who profoundly penetrated the phenomena of symmetry, called upon chemists to return to these ideas of Pasteur. His call met with only a weak response.

However, since then, the development of science has demanded the following of this path, to return to Pasteur and to P. Curie, who deepened Pasteur's ideas.

XII.

The phenomena of symmetry have not up until now been sufficiently embraced by philosophical and scientific thought. It is without any doubt the most fundamental and profound notion, which penetrates, in a subconscious way, our entire concept of the universe.

The revolution occurring in physics and the inevitable development of biological ideas tied to it, pose what seems to me to be the order of the day, the necessity of deepening and clarifying the principle of symmetry.

The most serious attempt to pursue the study of symmetry, although not completely followed through, was carried out by P. Curie, who considered symmetry fundamentally as a *state of space*, that is to say, as a structure of physical space.

This determination must be taken into consideration at the present time for the analysis of physical time, as in natural processes "space" and "time" are inseparable.

We can pursue the philosophical and mathematical analysis of the doctrine of symmetry more profoundly, but for our problem, and remaining within the empirical universe of the naturalist, this conception of symmetry, broad and real, is sufficient.

The phenomena of symmetry have, overall, drawn the attention of physicists only in the 20th Century when the enormous importance of crystallography with all its branches was definitively clarified in the domain of the physical sciences.

It was by way of crystallography and mineralogy that the doctrine of symmetry entered into physics. Even the most mathematical parts of this doctrine were elaborated with great precision and depth by mineralogists, who in this case always considered first their own problems, the problems of crystallography. Their acquisitions had been insufficient for physics, just as was proved by Pierre Curie.

They are insufficient in their current form, for the phenomena of life, which historically gave birth to the notion of symmetry itself, as this notion had its origin at the time of the work of sculptors who modeled living objects. The ancient Hellenes attributed the first formulation of the notion of symmetry in connection with the problem of the reproduction of the human body to Pythagoras of Rhegium, who lived more than 2,400 years ago.

And later, one of the founders of the doctrine of symmetry in mineralogy, A. Bravais, the eccentric French scientist, took symmetry, manifested in plants, as a point of departure for his work and created the doctrine of symmetry, basing his work simultaneously upon plants, minerals, and geometrical polyhedra.

But whereas the study of natural crystals blossomed in light of the doctrine of symmetry, the application of symmetry to living objects to which it owes its origin, and to physical phenomena, has always been sporadic and detached.

This had repercussions for the position of the doctrine of symmetry in contemporary scientific organization. The doctrine of symmetry is ordinarily connected to the teaching of mineralogy and the neighboring sciences, and does not hold the place which is due it, either in the discipline of physics, or of biology.

This is apparent in the lack of precision of representations of symmetry, which do not matter much for either crystallography or mineralogy, and in particular in the notion of dissymmetry, the importance of which was noted by L. Pasteur (for biology) and by P. Curie (for physics).

Pierre Curie (1858–1906), who continued the work of Pasteur in the study of symmetry, which he considered to be a "state of space" according to Vernadsky. "Curie's Principle," that "dissymmetry comes from dissymmetry," offers a new perspective on Redi's Principle, that "life comes from life."

XIII.

The term dissymmetry refers to diverse phenomena. For living bodies, for example, we can indicate two such phenomena which are demonstrated there simultaneously, but which are nonetheless independent. One of these phenomena is related to the doctrine of symmetry, whereas the other is not at all, but can only be studied on the basis of it.

In elaborating his great empirical generalization, Pasteur noted simultaneously the two phenomena in the state of space of living organisms.

At his time, even the notion of symmetry itself did not correspond to the current doctrine.

Although J. Hessel[18] had resolved, fifteen years before Pasteur, the problem of symmetry in a general form for crystals, his work did not draw much attention and did not enter life[19] for another thirty years, well after the discoveries of Pasteur. Pasteur had not yet reunited holohedry with hemihedry as we do today.[20] He did not realize that the optical and crystalline properties are always different manifestations of the same phenomenon—the phenom-

---

18. Johann Friedrich Christian Hessel M.D., Ph.D. (1796–1872), a German physician and professor of mineralogy who made significant contributions to the field of crystallography.

19. The meaning of the word life (*vie*) is unclear here, as far as whether it refers his ideas affecting the study of life per se, or the life of scientists more generally.

20. Holohedry and hemihedry relate to the symmetry of faces. In hemihedry there is, speaking generally, less symmetry, as for example, in the irregular shapes which are found on tartaric acid crystals, which Louis Pasteur studied. The faces of the tartaric acid crystals are said to be hemihedral.

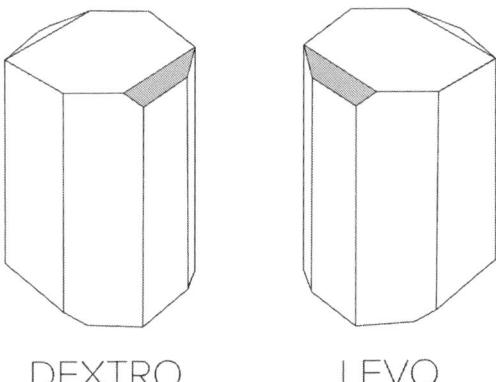

Examples of dextrotartaric and levotartaric acid crystals. Note that the truncated corners of the two different tartaric acid crystals (shaded) are what create their chiral character. Tartaric acid is a by-product of fermentation and can be found in the residues of wine.

enon of symmetry—just as we now accept it.[21] He found this connection in a particular case, and upon this basis, constructed his terminology, which did not enter later into common usage and which is rarely used even in his own country, in France. We find the same terminology in a more general form, in a more precise construction of Curie, which he does not indicate.

In studying the crystalline forms of organic compounds, existing in organisms or derived from them, Pasteur noticed the diminution of their symmetry, the appearance of forms—left and right—in the case where the racemic bodies were split up into their left and right antipodes. He called this phenomenon dissymmetry, that is, the violation of symmetry, as with regard to the polyhedra of racemic compounds the violation of symmetry was ex-

---

21. Right-handed tartaric acid is said to rotate light and be "hemihedral" in the same direction; its corners being truncated. See picture of two tartaric acid crystals.

pressed by the regular lack of either right or left faces of the antipodes.

He noticed that the polyhedra formed in this way were lacking centers and planes of symmetry whereas the fundamental polyhedra of racemic compounds, by the separation of which compounds the right and left antipodes had been obtained, possessed a center and planes of symmetry.[22]

He simultaneously proved that whereas the racemic polyhedra were optically inert in solution, the solutions of their antipodes rotated the planes of polarization—the right ones to the right, the left ones to the left.

He considered these two phenomena as a demonstration of the phenomenon of dissymmetry and as this demonstration remains stable in the liquid state, he named it *molecular dissymmetry*, seeking an explanation of the phenomenon in the structure of the chemical molecules.

Here, I cannot present the current composition of the phenomenon discovered by Pasteur. But it is important, nonetheless, to dwell upon it a bit.

Today we know that among the 32 classes of crystals, 13 correspond to the dissymmetry of Pasteur, that is to say that they possess no centers or planes of symmetry, but, with the exception of one single case, have axes of symmetry, which turn the planes of polarization to the right or left in determined vectors and which give right polyhedra in the first case and left in the second.

In addition, we know that these properties of crystals are expressed by the helicoidal distribution of their atoms—right and left—as required by the molecular dis-

---

22. Referring to the just cited image of tartaric acid crystals can help with envisioning this.

symmetry of Pasteur.

But this dissymmetry is only manifested in solutions, in liquids in the case where we observe, in the chemical structure, materials known to Pasteur, the so-called asymmetric carbon to which all the bonds are united to diverse atoms or groups of atoms. In the formulas of the chemists, the asymmetric carbon can actually lack even one element of symmetry in the surrounding space: it can be truly asymmetric. But the entire space of the molecule in which it is located will be dissymmetric, that is, it will possess some[23] axes of symmetry.

We are still awaiting more developments in the domain of the phenomena of symmetry. But at the same time, Pasteur discovered a new phenomenon while studying the phenomena of dissymmetry in relation to living matter, also related to the diminution of symmetry, that is to say the dissymmetry which, however, is located outside of the domain of the phenomena of symmetry and cannot be explained or predicted by it.

He discovered that in certain cases, instead of two antipodes, right and left, appearing simultaneously and in equal number, as required by the laws of symmetry, only one of the two antipodes emerges, or one of the two predominates clearly over the other.

As Pasteur did not, in general, know that one part of the violation of symmetry—which he called dissymmetry—could, actually, be deduced from the laws of symmetry, he did not distinguish this type of dissymmetry from other types which he had discovered, treating them as phenom-

---

23. Word "some" added here for clarity. The distinction is that something asymmetric will have no axes of symmetry, whereas something dissymmetric will have some axes of symmetry, as well as axes of dissymmetry.

ena of the same type; however, he noticed that the latter phenomenon was exclusively connected with life, whereas the former could be independent of it.

From the physical point of view, these two phenomena, both called dissymmetrical, are fundamentally distinct. The first is related to the distribution of objects in space, studied by the doctrine of symmetry. The second is not related to symmetry and is a real violation of it which cannot be predicted on the basis of symmetry.[24]

The principle of Curie, according to which every phenomena possessing dissymmetry must result from a cause possessing the same dissymmetry, is so general that it includes both phenomena.

---

24. It appears that Vernadsky is talking about the structure of the crystals as the first case, and secondly the optical rotation in solution. Handed crystals, such as quartz, exist, but do not exhibit the rotation of polarized light that solutions made from organic compounds do.

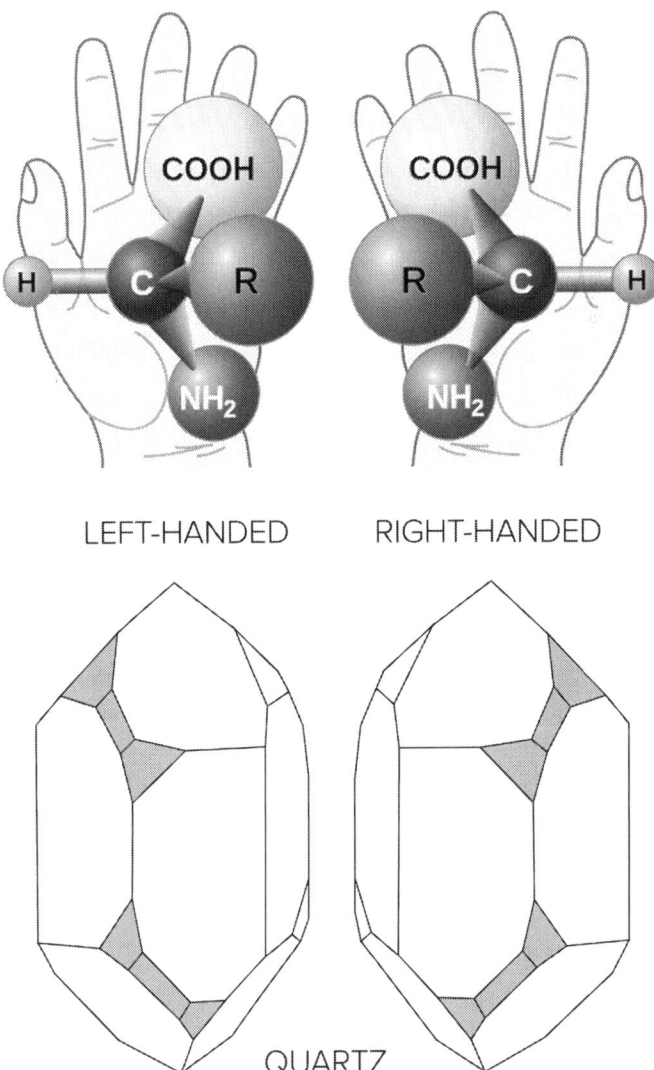

Above, a representation of the molecules of amino acids, and below, quartz crystals. Solutions of organic matter, from life, can rotate a plane of polarized light, but dissolved quartz, which is inorganic, does not. They are both dissymmetric, but not in the same way.

XIV.

Before expounding on the acquisitions of Pasteur, let us dwell upon the character of space which stems from symmetry, on its distinction from that of our space, the space of physics and geometry. It is precisely this specific space which we observe everywhere in organisms, conforming to the discovery of Pasteur and the principle of Curie—in the interior of bacteria or of the elephant for example. Certain properties of such a—let us call it—enantiomorphic space—right or left—must be apparent outside, in the surrounding environment of organisms during their life.

The distinction of such a space from regular space can be clearly expressed by the study of physical properties of vectors which are found there: that is, by the study of direction.

I already indicated that the phenomena of life are irreversible in time, that it to say, they always advance with the progress of time in one direction, without retracing their steps. The organism grows, ages, and ends in death.

There are no reversible phenomena: man has only imagined them in fairy tales and fantasies. In certain cases the signs of a reversible process can be observed as the late eminent Russian zoologist Schmidt[25] and recently C. Davidoff showed. But it is not these particular phenomena which characterize the life of the individual and the evolution of species.

Geometrically, the time of such a process can be can be expressed in the form of a vector AB, which is not identi-

---

25. Likely a reference to Peter (Petr) Schmidt, who appears to have been a zoologist and professor in St. Petersburg, in addition to possibly holding more posts.

cal to BA (−). The time of such a process is at least lacking a center of symmetry (the physicists sometimes incorrectly call it asymmetrical time). Whereas for the reversible process, AB=BA. The two vectors there are identical.

We can express this phenomenon by calling the first vectors *polar*, and the second *isotropic*. Time is geometrically expressed in the phenomena of life by polar vectors, and in regular phenomena—by isotropic vectors.

Space and time are inseparable in the new physics as well as in the real world of the naturalist. The ideas of Einstein are closer in this sense to the scientific concepts of the naturalist than the ideas of Newton, where time does not appear in the force of gravitation.

This explains the difficulty that the Newtonian theory has seen in penetrating the scientific environment, having required two to three generations to be accepted, as well as the celerity with which it has disappeared today from our field of view.[26,27]

---

26. [The brilliant and interesting lessons of M. Eddington on the nature of the physical world (1929) allow us, for example, to judge the depth with which the concept of the Universe of Newton has penetrated today, in its scientific part—the independence of space from time—into scientific opinion. In setting out the fundamental ideas of the new physics, M. Eddington based himself on the Universe of Einstein in which space and time are inseparable. And yet he admitted that the nature and the role of physical time, comparatively to that of physical space, was something else altogether. Recognizing, for the notion of time, a logical genesis of double nature—investigation of the external and internal experience of the living being (man)—he did not admit the same fact for space, not recognizing that these two phenomena were inseparable, according to the concept of the Universe of Einstein, and that these two were equally included in the particularities of the "space-time" of the living being. Incomprehensibly, he did not take into account the discovery of Pasteur, of the particular state of space of life.]

27. Based on a still in progress reading of the referenced book, *The*

The characteristic polar vectors for time therefore also have to characterize space, that is to say, the volume occupied by the bodies of organisms.

The phenomena of dissymmetry, characteristic, according to Pasteur, of these bodies, not only confirm this fact, but indicate again that these polar vectors must be *enantiomorphic*.

The direction AB is distinct from that of BA, but simultaneously, motion in the right and left directions around the vector in its surrounding environment can also be physically distinct. We distinguish the right and left vectors according to the helical direction of objects or motions in relation to the given vector. We distinguish, in this way, between four vectors upon one line:

AB (+)........... right and left;
BA (−)........... right and left.

In the case where certain single vectors—right or left—predominate in space, there we distinguish between two

---

Nature of the Physical World (1929), Eddington is at least in part referring to the debate initiated by Henri Bergson, whom Eddington refers to, regarding a concept of time which he called "la durée" or the psychological experience of time, which Bergson says is not touched by the theory of relativity. In the cited book, Eddington chooses to make the explicit point that there is no such psychological notion of space: "When I close my eyes and retreat into my inner mind, I feel myself enduring, I do not feel myself extensive. It is this feeling of time as affecting ourselves and not merely as existing in the relations of external events which is so peculiarly characteristic of it; space on the other hand is always appreciated as something external." It is not clear if Bergson made this point himself. Vernadsky referenced Bergson's concept of "durée" in "The Problem of Time in Contemporary Science." A book review by Vernadsky of Eddington's book exists in Russian. It appears that Vernadsky is taking issue with Eddington's statements implying that the main characteristic of space is simply extension.

distinct spaces, right and left. This is what Pasteur discovered for the phenomena of life.

We can and must go further.

The doctrine of symmetry includes a fundamental principle, indicating that the real structure of space where that structure appears is characterized by the minimum symmetry of phenomena observed there. It follows that there cannot be a center of symmetry in cosmic space, studied by physics, because otherwise we would not have observed polar vectors in one of its phenomena. But this space also cannot be characterized by planes of symmetry, as there would not then be enantiomorphic vectors in its other phenomenon, in the domain of life.

The space as well as the time of the old physics was isotropic: the vectors there corresponded, in their properties, to ordinary lines.

The space of the new physics is anisotropic. It can only include, in extreme cases, axes of symmetry. It is possible that this space is completely asymmetric, that is, that it does not possess any axes of symmetry. In this case, its

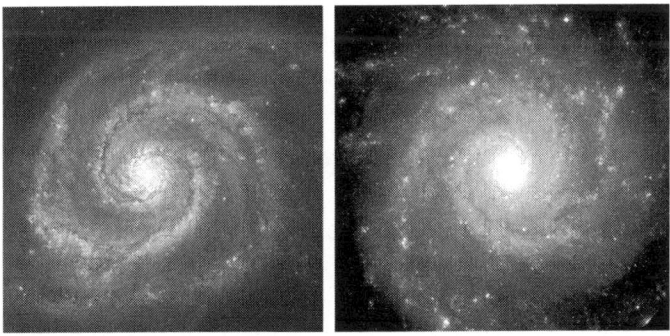

Vernadsky, as Louis Pasteur before him, hypothesized that there might be a real connection between the handedness of life on Earth, and manifestations of handedness in the cosmos, such as spiral galaxies, seen above in the cases of galaxies M51 and M74. Pasteur said, "I see dissymmetry everywhere in the universe."

properties, the properties of everything, will not be predicted by the doctrine of symmetry: all vectors will be polar, enantiomorphic and different in their numerical size.

The study of the physico-chemical properties of the field of life gives us, from this point of view, the most precise and profound indications, unlike any other phenomenon, meanwhile, of the Cosmos of physics.

XV.

Now let us focus our attention on the state of space embraced by life, as it appears to us in light of the discoveries of Pasteur, which remain until now the basis of our knowledge in this domain.

There exist a large number of biological observations related to the same domain which confirm the generalization of Pasteur, but they are disperse, not systematized, and not coordinated by thought in general.

I will return to this again; now, I take up the discoveries of Pasteur.

Pasteur unquestionably settled the matter of the dissymmetric structure—the absence of a center of symmetry and planes of symmetry—for all the main compounds produced by organisms and their products. The experience of more than half a century of biochemistry absolutely confirms this fact.

He named this dissymmetry molecular, as it is not only manifested in crystals, but in the liquid phase and in solutions. It is related to the helicoidal distribution of atoms in space, conforming to the laws of the symmetry of crystals. Albumins, fats, carbohydrates, the alkaloids, hydrocarbons, and sugars are all dissymmetric. All chemical bodies making up grains and eggs are without exception clearly dissymmetric.

Natural inorganic compounds, inorganic minerals, in no case manifest such a dissymmetry, that is to say that the property of rotating the plane of polarization of light in the liquid state or in solutions, is not present in them.

The deduction of Pasteur showing that molecular dissymmetry characterized the matter of living organisms and that it was not observed in the cosmic environment of surrounding life, remains unshakeable.

In this environment, we only know of petroleums which possess molecular dissymmetry and certain minerals with a helicoidal distribution of atoms in space (for example, quartz crystals). But the number of antipodes among the inorganic bodies of nature is never unequal. In the same deposits, we encounter right and left-handed quartz crystals of equal number. The contrary fact is observed for the compounds of living things.

At first glance, Pasteur had considered that the phenomena of life were distinguished from inorganic phenomena by their molecular dissymmetry, by their connection with the distribution of molecules (and respectively their atoms) in space. This distinction has disappeared for us; the dissymmetry of quartz is also connected to the distribution of atoms of silicon and oxygen in space.

Later, and to this day, the character of dissymmetry, discovered by Pasteur, was explained by the specific asymmetry of the carbon atom in the molecules of compounds, introduced by Le Bel and Van 't Hoff. But we are currently discovering other asymmetric atoms in molecules, such as Al, N, etc.

The phenomenon is probably connected to the stability of classes of symmetry in the solid state, without centers or planes of symmetry, for molecules having asymmetric atomic fields. This is only observed in nature in living organisms.

Pasteur deduced from this, with good reason, that such a clear difference between the matter of living organisms and abiotic matter had to be closely connected to the fundamental properties of the manifestation of life, and that it inevitably required particular cosmic forces by whose action life becomes manifest.

He said:

> If the immediate principles of life are dissymmetric, it is because, in their development, dissymmetric cosmic forces preside; it is there, according to me, that one of the connections between life on the surface of the Earth and the Cosmos, that is, the totality of forces spread throughout the universe, lies.[28]

And again:

> I see dissymmetry everywhere in the universe. Because we are coming to see that there had only been one single case where the right molecules differed from their left, the case where they are subjected to actions of a dissymmetric order. These dissymmetrical actions, possibly placed under cosmic influences, do these reside in light, electricity, magnetism, or heat? Would they be related to the motion of the Earth, with the electric currents by which physicists explain the terrestrial magnetic poles?[29]

> What could be the nature of these dissymmetric actions? I think, for my part, that they are of a cosmic order. The universe is a dissymmetrical unity, and I am convinced that life, such as it presents itself to us, is a function of the dissymmetry of the universe or of the consequences to which it leads. The motion of light from the sun is dissymmetric.[30]

---

28. [*Œuvres de Pasteur* (*Works of Pasteur*), Volume I, 373.] (Translator has located this quote on p. 375)
29. [*Works*, I, 361 (1860).] (Translator has located this quote on p. 341)
30. [*Works*, I, 341 (1860).] (Translator has located this quote on p. 361)

It is very characteristic that it is *one single antipode* which predominates or exists exclusively in the compounds connected with life. The other appears not at all, or only rarely, although it is possible to obtain it in the laboratory. I will note that following the principle of Curie, our chemical synthesis is provoked by a dissymmetrical cause expressed by the intelligence and the will of the experimenter.

Pasteur considered that only the right-handed forms of matter were stable in living organisms, that is, that the space occupied by life favors only the preservation of these molecular structures. He thought that we only observed right-handed antipodes in the most important matter of organisms—in seeds and eggs.

In short, the generalization of Pasteur, which unfortunately did not draw sufficient attention from the biochemists, holds true, although the right or left character of compounds is a more complex phenomenon than Pasteur believed.

The principal fact is the stability of one antipode in the field of life and the disappearance of the other. The predominance of the right antipode finds no current explanation; incidentally, the stability of a single antipode and not of other also has no explanation.

Pasteur was always concerned with this problem. He said:

> To understand the exclusive formation of molecules of a single order of dissymmetry it therefore suffices to admit that *at the moment of their grouping the elemental atoms are subjected to a dissymmetric influence* and as all other organic molecules which were given birth to in analogous circumstances are identical, whatever the origin and place of their production may be, *this influence*

should be universal. It encompasses the entire terrestrial globe.[31]

This phenomenon puts a very clear limit between the enantiomorphic forms created in the thermodynamic field of life and those of the surrounding cosmic environment where they are also located.

It is important to note that in the unique group of minerals characterized by molecular symmetry—in petroleums—we observe, first of all, their genesis by the metamorphosis of the remains of living matter, and second, the marked predominance of right-handed rotation in petroleums. Left-handed petroleums are very rare.

Ten years after his generalization, Pasteur went further and established a new fact in this domain which is no less important. It was in 1858, 71 years ago. He discovered that living organisms behave differently with right antipodes than with left. They can assimilate the right-handed antipodes and do not touch the left-handed ones. This is certainly a fact of great importance. According to Curie's principle, he established in this experimental way the dissymmetry of the living organism. Pasteur proved this for yeasts and for some molds. This was observed later for bacteria. This fact is thus established for the two forms of life, for life in the world of molecular phenomena and for that in our world of gravitation.

At first glance, this seems to explain the marked dominance of right antipodes in the production of life.

Yet, in reality this explains nothing; the fundamental problem remains unresolved—why do organisms only assimilate one antipode?

Why does the material of organisms allow right-handed

---

31. [*Works*, I, 241.]

antipodes to penetrate it and not allow left-handed[32] ones?

Taking symmetry as a starting point, Pasteur accepted the possibility of another life in another left-handed space with inverse antipodes—left-handed.

If the observed phenomenon is related to the state of space occupied by life, the right-handed space must, for for reasons currently incomprehensible to us, include the entire solar system, and possibly the galactic system.

Profoundly conscious of the immense impact of his discovery, Pasteur rightly affirmed that he had found an incontestable proof that:

> molecular dissymmetry, to date the exclusive privilege of products produced under the influence of life, appears as the modifier of physical and chemical phenomena peculiar to the organism.[33]

---

32. The translator takes some issue with Vernadsky's statement here, in that it conveys something seemingly absolute which Pasteur did not believe, that is, that life on earth had an exclusively right-handed character. Perhaps it is a mistake, and left and right were unintentionally switched in the text at this point, as today any statement about life's homochirality would state the opposite, that life on Earth has a left-handed character because of its amino acids, which are left-handed. Vernadsky himself said, in his paper "On the States of Physical Space": "Pasteur suggested that in some past period of geological history, *the Solar System had passed through left cosmic space* and that life had originated at that time, and reflected this phenomenon." A cursory reading of several of the cited passages from Pasteur's *Works* makes quite clear that he knew that amino acids, citing albumin specifically, are left-handed, and that sugars, such as tartaric acid which Pasteur famously and frequently studied, are right-handed. Were the text referring only to sugars, such as tartaric acid, it would be a true statement to say that life on Earth has an exclusively right-handed character. It is not known whether Vernadsky took a statement by Pasteur somewhat out of context here, in conveying a right-handed character to all living matter on Earth, or whether it might be a mistake.

33. [*Works*, II, 1922, 622 (1858).]

The ideas of Pasteur have received no response; the facts established by him were not developed.[34]

We have not advanced a single step in the course of these 80 years on the path cleared by Pasteur; we halt, powerless, before the enigmas he brought to light.

We have not done it, although their importance and the real possibility of studying them experimentally are clear.

This study is important not only for the most complete knowledge of life, but no less for the investigation of the state of physical space in general, because it reveals its new properties which do not appear in any other physical form.

Already, the unique capability of the living organism to distinguish these chemical and physical properties of the environment of life in their relationship with the enantiomorphic vectors is a phenomenon of exclusive importance.

The empirical generalization of Pasteur becomes very interesting today due to the creation of the new physics and the new picture of the Cosmos.

A great number of conclusions accessible to experiment stem from this, on which I cannot however dwell here.

It is important to emphasize the fundamental deduction: the phenomena of life allow us to push the study of the space of the Cosmos further than was possible in any

---

34. The following quote from Louis Pasteur very much conveys Vernadsky's message here, that the study of life necessitates a new physics: "You place matter before life and you decide that matter has existed for all eternity. How do you know that the incessant progress of science will not compel scientists to consider that life has existed during eternity, and not matter? You pass from matter to life because your intelligence of today cannot conceive things otherwise. How do you know that in ten thousand years, one will not consider it more likely that matter has emerged from life?" From *Pasteur et la philosophie*, Patrice Pinet, Editions L'Harmattan, p. 63, as cited in English at: http://en.wikiquote.org/wiki/Louis_Pasteur

other way. It is the cosmic nature of life which becomes apparent.

Pasteur saw this clearly.

XVI.

Numerous other (related) phenomena have been known in biology for a long time, but were unfortunately not collected and gathered together by systematic scientific thought.

One of these phenomena had already, at the end of the 18th Century, drawn the attention of a French writer and scientist, whose name was then famous: a writer who left a profound imprint on the sentiments and thoughts of men of the 18th Century, a precursor of romanticism on the stage of the last century, Bernardin St. Pierre. He wrote in his *Etudes de la nature* (Studies of Nature),

> It is very remarkable, for example, that all the oceans are full of univalve shells of an infinite number of very different species, which all have their spirals, which grow towards to the same side, that is to say, from left to right, as the motion of the earth (globe) when we turn the opening of the shell to the North and towards the Earth. There are only a small number of species that are exceptions and which, for this reason, we call unique. These forms go from right to left. A direction so general and exceptions which are so particular in the shells must, without a doubt, have their causes in nature and their age in unknown centuries when their germs were created.

Bernardin St. Pierre is more of an artist than a scientist and that being the case, he often embraced, with reason and according to his cosmic sentiment of nature, the great phenomenon of life which the experimenter Pasteur approached fifty years after him.

Here, we are approaching an immense domain of facts not yet affected by exact scientific thought.

It is necessary, from now on, to put forward the most important indications which evoke our curiosity. I can only make brief remarks about them here. So first of all, it seems that the directions of seashell spirals of the same species can change in the course of geological time. There does exist, for example, an indication that the shells of all *Fusus antiquus* of the inferior red sandstone of England (Lower Permian) are all left-handed, whereas the modern ones are all right-handed. If there were no cause—necessarily dissymmetric, according to the principle of Curie—upsetting the symmetry, we would have an equal number of right and left-handed spirals. The cause which determined this phenomenon was thus changed in the course of geological time. It was left enantiomorphic in the given location during the Permian Epoch, and right enantiomorphic during our time.

Shell of the Fusus antiquus

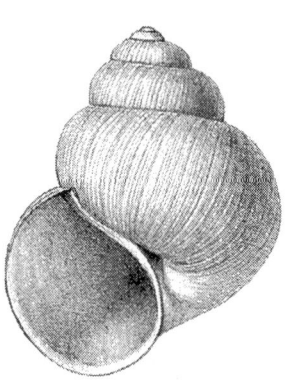

Drawing of the shell of the Lanistes ovum, *a species of African freshwater snail*

The fact that the embryos of mollusks give, in a number of cases, left-handed spirals, whereas the adult forms give right-handed ones, indicates, it seems, the possibility of such a change of the space of life.

We stop here, while waiting, powerless, in the face of a need for an explanation of this phenomenon. It is important, above all, to study it and to confirm it. The phenomenon is certainly very complex. Thus, today there are also species of mollusks with left-handed spirals, although their relative number is small when we study them all together.

Moreover, geographic changes are brought to our attention: the *Lanistes* of Lake Tanganyika has left-handed spirals and the same genus, living in the neighboring lakes of Nyassa and Victoria have right-handed spirals. What is the cause of this phenomenon?[35]

Innumerable observations of the same type, gathered together, are scattered throughout the scientific literature, on other spirals of plants and animals which are found everywhere—forms of seeds, of flowers, etc. Clearly, here we find ourselves in the domain of dissymmetrical phenomena, closely connected with the problems treated by Pasteur, but which are not at all touched upon by theoretical thought.

It is not impossible that in studying these, we will find specific properties of the space connected with life or with unknown dissymmetrical forms.

The work of our current time and of the near future demands that we follow the paths which open up to us.

---

35. Lakes Tanganyika, Nysassa (also known as Lake Malawi), and Victoria are all located in eastern Africa, in the regional vicinity of present-day Tanzania. Freshwater snails of the genus *Lanistes* are found in all three lakes, however, as Vernadsky observes here, the shells of those in Lake Tanganyika exhibit left-handed spiral formation, while those in Lakes Nyassa and Victoria are right-handed.

XVII.

It seems that it is possible to study physical time no less profoundly by research into living phenomena.

The time of the physicist is certainly not the abstract time of the mathematician or the philosopher. Time is manifested in different phenomena under forms which are so different that we had to give it different names in our empirical science. We speak of historical, geological, and cosmic time, etc.

It is convenient to distinguish biological time by the limits in which living phenomena are manifest.

This biological time is now estimated at $2-3 \times 10^9$ years—by billions of years in the course of which the presence of biological processes known to us in the Cosmos began in the Archaezoic. It is very likely that these years only corresponded to the existence of our planet and not to the duration of life in the Cosmos. Today we arrive at the conclusion that the duration of the existence of celestial bodies in the Cosmos is also limited, that is, that we are, in that case, also dealing with an irreversible process. We ignore the duration of the manifestation of life in the Cosmos, our knowledge of life in the Cosmos being in general minimal. It is possible that billions of years make up only a very small part of biological time.

For life on Earth, the irreversible process is expressed within the limits of this time by the evolution of species.

From the point of view of time, it is probably a manifestation of Redi's principle, that is to say the succession of generations, which must be considered as a fundamental phenomenon.

We have a number of phenomena in this succession of generations accessible to quantitative study and which

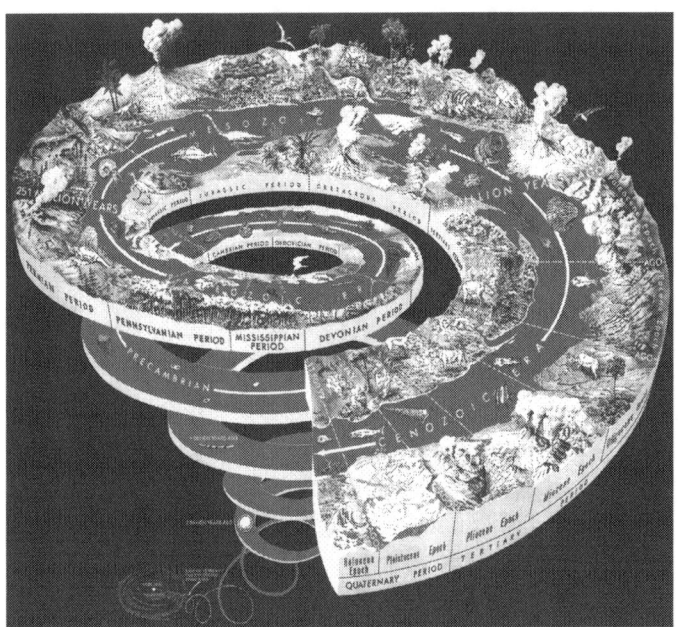

Depicted here is a model of the geological time scale with the evolution of life on Earth divided into its various eras and periods. Vernadsky saw the consistent upward evolution of life as a unique characteristic of biological time.

give an exact, mathematical and quantitative representation of the structure of the polar vector, which corresponds geometrically to the processes of evolution.

Unfortunately, the scientific facts related to it are dispersed and not always exact. Today, we can only evaluate the constants of biological time by the limits of numbers and not by the numbers themselves. But the change in our ideas about the position of life in the Cosmos urgently requires the organization of systematic experimental investigations in this direction.

The irrefutable existence of *a minimum limit of the duration of the succession of generations* is quite obvious. This limit indicates the minimum time necessary for the formation of a determined number of organisms, that is, not only for the formation of their mechanisms, but also for all their most complex chemical structures—albumins, etc. This phenomenon is clearly submitted to determined laws.

I tried elsewhere to establish that this limit corresponds to the duration of the minimum average division of the unicellular organism and is carried out with an intensity reaching the limit of what is physically possible.

The limit is not placed there by the short duration of the succession of generations, insufficient for the formation of innumerable and complex chemical compositions necessary for life, but by the properties of the physical environment and moreover by the properties of gas, by the respiration of organisms. The organism must carry out its gaseous exchange in such a way that its living environment is not destroyed. Thus the speed of the propagation of its geochemical energy by reproduction (the succession of generations) cannot exceed the speed of the sound wave in a gaseous medium, in which the organism breathes.

The fact that life can actually reach this limit proves the extreme intensity of the living process which is clearly not solely connected with the properties of the material medium.

The research of this limit is on the agenda. As far as we can judge, the minimum duration for the succession of generations is somewhere between 16 and 22 minutes, closer to 20 minutes it seems. This length requires an exact determination. It is an important biological constant. It can play the role of a natural unit for the study of biological time. We can consider it as a measure of biological time. Its determination does not seem to pose experimental difficulty.

It appears that there is also a maximum limit for the succession of generations. In some vegetable organisms, we observe that it corresponds to several hundreds of years, that is to say 10^7, perhaps 10^8 minutes. Its determination is also a matter of time.

Thus the amplitude of the variations of the succession of generations is very significant and can vary by millions or tens of millions.

The change of the duration of generations in the process of evolution, in the course of geological time, is very characteristic of biological time. We will not have an idea of this process and its character until there is a concentration of a sufficient quantity of facts. For man, the duration of a generation in the process of evolution seems to grow in the course of time.

The phenomenon must be studied on the basis of the new physics in the complex "Space-Time." The Space of life has, as we have seen, a particular symmetrical state which is unique in nature. The time which corresponds to it not only has the character of polar vectors, but a particular parameter, proper to itself, a particular unit of measure, connected with life.

I cannot dwell any longer on these phenomena. It is only important to me to make their importance known.

A multitude of problems suddenly appear; the possibility of a quantitative scientific investigation is clear.

It is not until after the facts which have been known for a long time become systematized or that new facts become amassed that we will be able to realize what this will contribute to the study of biological time in connection with the succession of living generations which characterizes it.

XVIII.

But it is clear, from the standpoint of the problem which interests us here,—that of the importance of the investigation of life for the construction of the scientific picture of the Universe,—that this investigation is of interest for the space and time of the Universe. It introduces new traits, not known from other physical or chemical phenomena.

It is clear that life cannot be separated from the Cosmos, and that its study must have an impact—perhaps very significant—on scientific representation. This does not only concern space and time, but also other fundamental elements of the Cosmos. I can only indicate them here.

Thus, life is set almost entirely apart from other phenomena in regards to the energetics in the Universe, in diminishing and never increasing its entropy. According to the opinion of Prof. Jaeger, life creates, through the evolutionary process, forms which are increasingly lacking in elements of symmetry. Finally, the intelligence of man begins to manifest itself today in the process of the biosphere, always more clearly and decisively, and changes the established geological process in a radical manner.

The new representations of the Universe created by the new physics compels us to pay special attention to the study of the phenomena of life which indicate their character which is not only terrestrial, but *cosmic*.

It is especially important because of the biological problems which suddenly arise, and which can be encompassed by number and measurement, the fundamental approach, leading to the construction of the scientific Universe.

Vast new horizons of research thus open themselves up

to biology. The scientific confirmation of the fact that life is not a planetary phenomenon, but rather cosmic, will have immense consequences for biological and humanitarian conceptions.

The future will decide whether this is the case. But as we wait, the development of the new physics allows us to follow not the way of philosophical constructions, always insufficient and precarious, but that of exact scientific research, based on number and measurement. The new way which has been cleared before us, will lead us, perhaps far from the biosphere, in which today the entire work of the biologist, and to a lesser degree that of the geochemist, is concentrated.

V. Vernadsky
Member of the Academy of Sciences of Leningrad
Correspondent of the Institut de France

21ˢᵗ Century Science & Technology
Special Anthology Series
150 Years of Vernadsky

Volume 1: The Biosphere
Volume 2: The Noösphere

A special two-volume anthology published in celebration of the sesquicentennial of the birth of Vladimir Vernadsky, featuring select original English translations of his writings, plus critical articles elaborating continuing research into the implications of his work concerning mankind's understanding of the universe we inhabit and our role within it as a uniquely noetic species.

http://bit.ly/vernadsky-150

Made in the USA
Lexington, KY
11 March 2017